Social Anxiety Explained

Forgot About Shyness, Anxiety and Improve Your Self-Esteem. Learn How to Eliminate Distorted Thoughts and How to Change Them.

George Kewell

The information in the following pages is broadly considered to be a truthful and accurate account of facts, and as such any inattention, use or misuse of the information in question by the reader will render any resulting actions solely under their purview. There are no scenarios in which the publisher or the original author of this work can be in any fashion deemed liable for any hardship or damages that may befall them after undertaking information described herein.

Additionally, the information in the following pages is intended only for informational purposes and should thus be thought of as universal. As befitting its nature, it is presented without assurance regarding its prolonged validity or interim quality. Trademarks that are mentioned are done without written consent and can in no way be considered an endorsement from the trademark holder.

Table of Contents

Introduction

Congratulations on purchasing *Social Anxiety and Shyness.* We will discuss ways to overcome social anxiety and shyness. Understanding the fundamental differences is important to work towards a solution. Hopefully, this book offers real solutions to problems with social anxiety that you may be experiencing.

The following chapters will discuss:

- What social anxiety is
- How to spot the difference between social anxiety and shyness
- Steps to overcome shyness
- How to identify social anxiety disorder
- Learn to spot the triggers and responses to social anxiety
- Find out ways to change your thought processes
- Learn strategies to switch from a negative mindset to a positive mindset
- Find new coping strategies to overcome social anxiety
- Positive affirmations to keep you focused
- Follow a 6-step process to be rid of social anxiety for good
- And more!

Social anxiety is a real disorder that should be dealt with as soon as possible. Hopefully, this book gives you solutions, or at the minimum, a place to start. Good luck on your journey, and we wish you much success!

There are plenty of books on this subject on the market, thanks again for choosing this one! Every effort was made to ensure it is full of as much useful information as possible, and please enjoy it!

Chapter 1: Social Anxiety

Social anxiety is when an individual fears social situations that involve any interaction with other individuals. This fear and anxiety cause the individual to feel judged negatively and also to be negatively evaluated by others.

This disorder is pervasive and causes anxiety and fear in just about every aspect of an individual's life. Social anxiety is common, and most people do not realize that there is a help. If you suffer from social anxiety or social phobia, you should know that there is hope for the future and that you do not have to suffer from this forever.

As a psychological disorder, social anxiety is recognized and treated by mental health individuals. Treatment options vary but do not wait to seek help because the longer you wait, the worse your symptoms can become, and the harder this disorder is to treat.

There is no shame with this disorder.

What Causes Social Anxiety?

An individual can be predisposed to having a social anxiety disorder if they have a parent or parents who also suffer from this. They have a 30 to 40% more likelihood of developing social anxiety themselves if their parents have suffered as well. There are links between hyperactivity and a part of your brain called the amygdala. Just this is where your flight or fight response is located. When this portion of your brain is triggered, an individual can become slammed with intense anxiety. Physiologically they will experience a racing heart, muscle tightening, and blood sugar levels will rise, excessive sweating, hyper respiration, and the freeze-up of the brain.

The freezing of the brain causes these individuals were suffering from anxiety to be unable to reason or think normally. These individuals also find that their mental focus shifts. In the prefrontal cortex of the brain is where your rationality comes into play, and the signals between the two portions of the brain become mixed up, and the prefrontal cortex is no longer able to calm you.

When you suffer from social anxiety, rather than allowing yourself to become calm, it instead amplifies the activity in the other portion of the brain.

There is hope, though because, with the use of cognitive-behavioral therapy, your brain can be reprogrammed and rewired. Besides being physiological, social anxiety can also be environmental.

When an individual has parents, who are critical, overly controlling, less affectionate, the child or individual will have a warped impression of the world that is going to be shaped by the characteristics of the parent.

As a result, these children and young people are more fearful and often trust others less if they, in this type of environment. The self-esteem and self-confidence are going to be negatively impacted as well. In most cases, the parents are not even realizing the damaging effect that their actions are having on this child.

Social anxiety disorders are often not diagnosed until adulthood, but the symptoms tend to start showing up later in childhood and even in early adolescence.
Other environmental influences or harmful life experiences can also cause social anxiety. If an individual experiences stressful life events or dramatic experience during it's their childhood, it's going to influence their development and can cause social anxiety problems when looking at severe social anxiety.

Some dramatic events can include:

- physical, sexual, or emotional abuse
- Bullying or teasing by peers
- Family conflicts
- Domestic violence
- Divorce
- Death of a parent
- Desertion by a parent
- stress during the pregnancy or infancy

With traumatic experiences, these often lead to reinforcing in the individual that the world is unpredictable and terrifying. By experiencing the different life events that lead to trauma, the world has been identified as a place where horrible things happen. This leaves the individual feeling unsure and uncomfortable in the areas of life that are needed to thrive.

An interesting thought to remember is that not everyone with social anxiety has suffered a traumatic event. Your experience with social anxiety may look different than someone else's journey. The important part is, to begin with solid, effective techniques, as you will find in this book, that offers hope to the anxiety you are feeling. Not every one of these strategies will work for you, but finding a balance that is tailored to your needs is important.

What Is Shyness?

Shyness is the feeling where you an individual feel apprehensive, uncomfortable, her awkward. This is especially true when this individual is in unfamiliar surroundings, new situation, or with unfamiliar people.

A common characteristic of individuals who suffer from shyness is low self-esteem. Shyness should not be mistaken for social anxiety because they are different.

How Are Social Anxiety And Shyness Different?

Often social anxiety and shyness are mistaken and used interchangeably. Because of this, individuals are often told that they will outgrow their anxiety, or that they need to get over it. Shyness is something that an individual will gradually overcome while they become more familiar with their surroundings.

Social anxiety is a mental health disorder that requires professional help, coping strategies, and improved thought processes.

- Shyness may disappear when a person becomes more familiar with their surroundings.

- Shy people tend to be more optimistic.

- Individuals who suffer from shyness is not self-critical, or at least not as self-critical as those suffering from social anxiety.

- Shyness does not require medical intervention.

Seven ways to overcome shyness

If you have determined after looking at the signs and symptoms of both social anxiety and shyness, and you have determined that you will more than likely suffer from shyness, below you will find some tips to overcome this feeling.

Act With Confidence

When you act confidently, you are going to feel more confident and bolder. When you are trying to master confidence, it is okay to simply fake it until you make it. Confidence needs to be developed, just like any other skill.

Engage With Others

If you are suffering from shyness, it's a good idea to find others to engage with. It's easier to overcome the feelings of anxiety or trepidation that you may experience with shyness if you familiarize yourself with different situations.

Understand that as you become more familiar with a vast amount of other people, you are going to feel more comfortable you are going to feel more confident, and you are going to take those feelings and overcome your shyness.

Even If You're Anxious, Try New Things

It's always important to expand your comfort zone, but when you suffer from shyness, this takes on a whole new meaning. Allow yourself to experience new ideas, people, places, and things. By doing this regularly, you are going to find the shyness that you feel lasts less and less.

Talk About Your Insecurity And Nervousness

When you suffer from shyness, it can be difficult to talk to others about what you are experiencing, but you must talk just the same. The more you rehash what you are feeling nervous about, the less hold it has over you. Remember, the end goal is to overcome your shyness. Sometimes you need to be uncomfortable in the process.

Make Yourself Vulnerable To Others

Making yourself vulnerable is going to be when you talk to others about what you are experiencing. Admitting that you are a bit nervous in situations is okay and often can lead to a way to break the ice. Knowing that others are more than likely going to be understanding in the situation should help you. Rarely everyone is completely confident and free of nerves in new situations.

Practice Confident Body Language

When you're mastering your shyness, you must work hard to portray confident body language. Leave the slouching at home. Stop avoiding eye contact, and stand up straight and tall. When you display this type of body language you were going to find that you feel more confident each time you are put into a different situation

Practice Mindfulness

Whether you're suffering from social anxiety or shyness, mindfulness is a good way to ensure that you remain in the present and stop worrying about the past and the future. Practicing mindfulness is a good technique that also lets you enjoy the moment. Life is about living in the moment and taking full advantage of each moment you are given. Take advantage of that.

How do You Practice Mindfulness?

Remember that mindfulness is simply keeping yourself grounded and rooted in the present moment, releasing the past and future. This is a skill that you can develop to help you overcome worry, anxiety, and stress.
Being mindful allows you to keep control of your thoughts, emotions, and feelings. But how do you start practicing mindfulness?

Meditate
Meditation is a great way to get started practicing mindfulness. A bonus is that you can also use meditation to get a handle on anxiety and negative thought processes. Your mind is allowed to explore and wander about without distraction. You can spark your natural curiosity through meditation.

To do this, you need to suspend the judgment you may be feeling about this sort of exercise. Mediation used to have a negative stigma, but no more!

People from all walks of life are using meditation to conquer their goals, find their path, and overcome crippling anxiety.

Steps to Practicing Mindfulness

1. Make time-to start practicing mindfulness, and you need to carve out time in your schedule to specifically focus on this task. You need not have special equipment, only in space and time. Dedicate a small patch of realty in your home to practice your mindfulness exercises.

2. Take in the present moment-By looking at the reality of the moment, and you are aiming to achieve inner calm. When you pay attention, refrain from judgment. Take at the moment as it is happening. Don't worry, and this is difficult for everyone at first!

3. Ignore Judgement-When you feel like you are going to pass judgment, urge that feeling and thought by and ignore it.

4. Return to the present-Find your way back to the present moment and experience it again without judgment. Remember that you are in control.

5. Don't judge the thoughts when your mind wanders, do not judge the strange ideas that may pop up.

Give yourself the allowance to ignore the judgment, and have fun with what happens!

The important idea here to remember is to practice. You must practice this to use it during stressful times and situations. If you do not practice, the exercises will not be effective. Compare it to never practicing push-ups, but then going to a competition and demanding your body do 150 push=ups perfectly. Sounds silly, right? Don't expect the same results with mental exercises. Practice always makes perfect, and you need to aim for perfection.

Another good option when practicing mindfulness and relieve from anxiety is to meditate.

Below you will find the basics of meditation. Remember that you can always seek out guided meditations that take you on a journey and gives you the directions.

1. Find a comfortable spot to sit. You need to be stable, so try the floor with a cushion perhaps.

2. Pay attention to your legs. If you are sitting on the floor, cross the legs comfortably. On a chair, keep your feet planted firmly on the ground.

3. Keep your upper body straight, but not stiff.

4. Notice your arms. Are they parallel to your body? If not, position them in a way that feels natural. Rest your palms on your legs.

5. Drop your gaze. Allow your chin to dip down and point towards the floor. By doing this, you are allowing your eyes to see what is there without having to focus too hard.

6. Pay attention to your breathing. If you are feeling the sensation of your breath, then you are doing it right. Breath from your diaphragm, and allow your belly to rise and fall.

7. When you breathe, pay attention to where your mind wanders. Let your mind have free reign to think about topics that it wants.

8. Don't judge these thoughts. Let them come naturally.

9. Gently lift your gaze when you feel ready. If you have closed your eyes, now is the time to open them. Take time to notice your environment. Keep being aware of how your body feels in that moment. Pay attention to your thoughts and emotions.

Before you can use meditation as an effective calming tool, you need to practice. As with mindfulness, you need to pay attention and remember how the end feels. Consider journaling about your experience during meditation.

Benefits of Meditation and Mindfulness

The benefits are extras with meditation and mindfulness. Overcoming shyness will take some work, but you can do it. Have faith in yourself and tell yourself that you are capable of ridding yourself of the anxious feelings that you have in new situations. Remember, always ask for help if you need help. Social anxiety, on the other hand, is a different matter.

- Social anxiety can either remain unchanged or get worse despite becoming familiar with your surroundings.

- Socially anxious individuals are commonly pessimistic and tend to over-analyze simple situations.

- Social anxiety is often mistaken for extreme shyness

- Social anxiety disorder or sad is a medical condition, is recognized as a psychiatric condition, and frequently requires medical intervention.

- The symptoms of social anxiety do not go away, but often they become worse progressively.
- Individuals who suffer from social anxiety believe that everyone sees them unfavorably.
- These individuals put a lot of effort into avoiding social situations at any cost.
- The debilitating nature of this condition affects not only the social aspect of life but even how somebody functions.
- Social anxiety interferes with each aspect of an individual's life.
- Social anxiety can occur with other psychiatric conditions such as depression, other anxiety disorders, and substance abuse.

Is There Hope?

If, by now, you are beginning to think there is no hope for overcoming social anxiety, keep reading! Hope is located within these pages, and you can begin today! Most individuals do not seek help for their social anxiety disorder, which causes the disorder to become worse over time. You have already reached out in faith to look for solutions.

Keep reading to get a glimpse of the solutions and hope available to you in the upcoming chapters.

Cognitive-Behavioral Therapy

Cognitive-behavioral therapy or CBT is a process of working with a mental health professional to get a handle on the anxiety that an individual feels. This type of treatment is often extremely effective, and when coupled with other methods such as medication or self-help strategies, an individual can overcome the social anxiety that they are experiencing. There are three areas that CBT therapy should include.

Ongoing assessments to determine the level of anxiety

When you are engaging in cognitive behavioral therapy, you must do an initial assessment to mark the level of anxiety that you are feeling. From then on, it's important to periodically repeat these assessments to gauge how effective the current treatment is working. When you discuss this with your mental health professional, it's a good idea to be willing to take these assessments and encourage the mental health professional to take this path.

Knowing where you stand and giving a number to your current level is important. These assessments are a good way to gauge the progress that you've made.

Education

CBT should include educating the individual about anxiety. This education should be specific to the needs of each individual. When you are equipped to handle the anxiety that you are experiencing, it's going to benefit you in the long run. Having information and facts about a social anxiety disorder is going to give you a solid footing to know how to combat this behavior. The mental health professional can give you plenty of information. One of the benefits of being educated about your disorder is that you will better be able to self-advocate. Often individuals who suffer from social anxiety disorder do not advocate for themselves and tend to be passive rather than assertive.

Strategy Training

CBT should include solid backing that involves strategy training. This training should give you a strong self-help support system. Training in mindfulness, management of anxiety systems, and encouragement are important. Later on, in this book, we will talk about ways to practice mindfulness, and we will also discuss ways to manage anxiety symptoms. Having someone to encourage you along the way is going to benefit you by giving you support. In addition to a CBT professional, it would be a good idea also to have an accountability partner.

Behavioral Interventions

CBT should include behavior interventions, as well. Not only are you trying to treat the physical symptoms in the mentalist symptoms of social anxiety disorder, but you are also trying to break behavior patterns. By breaking behavior patterns that have allowed an individual to fall deeper and deeper into a rut, you are empowering the individual to take actions to correct these behaviors themselves. This empowerment is going to lead to greater satisfaction and help these individuals better themselves.

Focus On Gradual Integration

Exposure to fearful situations is important but should not be sudden. Diving headfirst into a new, fearful situation will not be effective for individuals suffering from a social anxiety disorder. These individuals need to have gradual exposure that changes slightly over time.

This does not mean that the behavioral intervention needs to be long and drawn out, but rather it's a good idea to allow the individual to adapt to their surroundings before introducing new stimuli.

Change The Situation

When you adopt that experiences to target slightly different problems or goals, you are helping the individual who suffers from social anxiety get a better handle on plans, not going the way they expected. Each adaptation should accommodate the specific goal or problem of these individuals have. You can include a familiar face, and next time have that phase be absent. Give the different individual scenarios to work with to better build a repertoire of skills. By giving these individuals who suffer from social anxiety the chance to work in a comfortable environment on their goals, and with support, you are giving them a greater chance for success.

Tackle The Unexpected

Social anxiety tricks your mind into thinking that every situation is the worst possible situation that could happen. When working on behavioral interventions, it's important to include scenarios that allow the individual to work on unexpected changes. This is where the individual needs to ask themselves what's the worst that could happen? From there, these individuals need to identify what sorts of catastrophes could happen. Once they have made their list, it's important to go through and determine the likelihood of each scenario. By identifying how likely a scenario is, it's going to be easier for these individuals to break a distorted thought process.

Identify Unhealthy Coping Mechanisms

Each individual with social anxiety has coping mechanisms that they have fine-tuned throughout their life. These coping mechanisms could be avoidance, isolation, and more. Helping the individual determine what coping mechanisms they use that are helpful and which are harmful is going to allow them the chance to make goals to tackle the specific instances. Through the help of behavioral interventions, individual suffering from social anxiety disorder will have the chance to work on healthy coping mechanisms such as deep breathing and meditation and get rid of the on healthy coping mechanisms that keep them stuck.

Cognitive Interventions

When you take part in cognitive behavioral therapy, not only are you tackling the behaviors that you are displaying, you're also tackling the thought processes involved. The idea is to help you change distorted thoughts and identify ways to rethink the situation.

Identify Your Fears And Challenges

The fears and challenges of an individual who suffers from social anxiety need to be identified before they can be worked on. By identifying the specific fears and challenges, the individual is going to have the opportunity to work on the specific thought distortion.

Labeling These Thought Distortions

By keeping a thought journal, you will be able to specifically pinpoint the different distorted thoughts you're having throughout your day. By labeling these distortions with specific situations that you encounter, you are better equipped to handle those situations at the moment. Breaking the negative thought patterns will be easier once you have made a personal connection.

Shifting Attention

An effective method to cope with social anxiety is by shifting your attention. This works well with CBT therapy because part of shifting your attention is breaking the negative and distorted thought process. Once you identify this thought process, you would need to practice mindfulness. Bringing yourself back to the moment is a good strategy to avoid a panic attack.

Medication

There are many different options for social anxiety with medication. Some may not be helpful because they allow a person to become tolerant and dependent on these medications. Selective serotonin reuptake inhibitors or SSRIs could be effective when used in combination with a CBT regimen. When you couple the use of medication with CBT and positive coping skills, the success rate for overcoming social anxiety rises dramatically.

There is hope for overcoming social anxiety and shyness. Depending on which one you suffer from, you have different available strategies. Depending on the approach that you take, social anxiety is treatable. Throughout this book, we will discuss strategies to overcome your social anxiety. The important idea to remember is that this takes time and that no matter what setbacks you may experience, your battle with social anxiety is fully treatable.

In this book, we are going to discuss strategies that you can use to alleviate your symptoms and overcome the crippling aspects of social anxiety. You will learn the tools to survive each area of life until you eventually thrive. The process will take time, but in the end, it will be worth it!

Chapter 2: Why Me?

As an individual who is suffering from social anxiety, you may be asking why has this happened to me? Are there others who suffer as much as I do? You may feel hopeless and distracted times because of the symptoms that are associated with your social anxiety. The chances are that if you have picked up this book, you understand that you may be suffering from social anxiety, but your whole life you might've been told that you are just extremely shy. Now that you understand the difference between shyness and social anxiety, you should be able to find new hope and begin the battle of overcoming.

Statistically, social anxiety is not uncommon. More individuals suffer from this, then even they realize. Some statistics include:

- 15 million adults suffer from this disorder making it the second most common diagnosis in the United States

- Symptoms typically begin around the age of 13, but they can begin at any age. The teen years and early adulthood are the most vulnerable ages.

- It typically takes ten years for an individual seek treatment

- There are nine common physical symptoms, but more are prevalent as well.

- 66% of individuals diagnosed with a social anxiety disorder also have another form of mental health disorder.

- In as little as five months, with proper treatment, you can begin to see improvement. This is going to vary among individuals, but know that it is possible.

Common Symptoms

The emotional symptoms of social anxiety disorder are going to vary by intensity depending on the individual. Keep in mind that the symptoms are common threads among many individuals but having them all is not a requirement. If you suffer from the majority of the symptoms, it's safe to assume that you are also suffering from a social anxiety disorder. If you're still unsure after looking at this list of symptoms, you can search online and find free evaluations that will tell you if your symptoms are from social anxiety or not.

Intense Fear Of Rejection, Criticism, Judgment, And Being Perceived Unfavorably

Most people think that individuals with social anxiety are overstressing their emotions. It is unclear to individuals who have not experienced these emotions to understand the debilitating nature. Well-meaning individuals may tell SAD (social anxiety disorder) sufferers to merely "get over it." While this is usually well-intended advice, for SAD sufferers, it is impossible. Every day of their lives is spent in fear of what will happen to make them criticized, rejected, or judged. This is a real fear of the sufferer, even if others do not understand. It is not something that is "all in their head" because physically, they are manifesting this fear. While it may be irrational fear and a product of distorted thinking, at the moment SAD sufferers are terrified.

Anxiety

Anxiety is the cornerstone of social anxiety disorder. The feelings of anxiety will limit the interactions of social anxiety sufferers. This feeling will keep them from doing the things that they love, keep them from enjoying life. The debilitating effect of the anxiety will cause changes to a person's personality, routine, and life.

Excessively High Level Of Fear

Fear is a common emotion in life. The difference between normal fear and the fear of a social anxiety disorder sufferers are not in the same ballpark. Normal, healthy fear keeps us from engaging in dangerous, life-threatening situations.The high level of fear that is experienced in a SAD sufferer is distorted. These individuals will experience life-threatening fear over common, everyday experiences.

Nervousness

Do you experience nervousness? Normal nervousness dissipates rather quickly, or quicker than a SAD sufferer. The nervousness experienced by a SAD sufferer permeates each aspect of their lives and causes them to anticipate failure constantly-even when failure is not possible.

Blushing

Some situations cause a person to blush: attention from a love interest, a faux pas, or an innocent mistake. Individuals who suffer from social anxiety will experience blushing despite the moment being benign.

Trembling

The tremors and the trembling that an individual who has social anxiety is a whole-body experience. They tremble at situations that others would not bat an eye at because of the symptoms of fear they experience.

Muscle Twitches

Muscle spasms and twitches are common manifestations of fear. For SAD sufferers, this is a common experience that they cannot control.

Racing Heart

During a fight or flight response situation, the heart rate increases, and this causes the individual to respond. If you are experiencing these symptoms during a banal situation, then you understand the experiences of a SAD sufferer.

Shortness Of Breath

The inability to breathe can be caused by any number of things: exercise, medical conditions, or panic attacks. For SAD sufferers, shortness of breath, and the inability to breath normally is a reaction to the fear they are experiencing.

Sweating

If you have been nervous, you understand that sweating is a reaction to that nervousness. When coupled with other symptoms, sweating becomes another unbearable symptom of social anxiety.

Your Mind Going Blank

Losing your train of thought can happen to anyone, but for SAD sufferers, this is a debilitating side-effect of social anxiety. Their inability to maintain their train of thought is not only unbearable, but it causes them intense feelings of embarrassment.

A Shaky, Soft Voice

When you try not to draw attention to yourself, a soft voice is key. Individuals with a social anxiety disorder will often speak softly to avoid attention. Their voice may be shaky or unsteady, which reflects their low confidence in their message.

Concentration Problems

Losing focus because of social anxiety is common. Individuals spend much of their time rehearsing what they are going to say and do the next time they are required to speak; thus, they can't focus on the task at hand.

Urge To Use The Toilet

Coupled with other symptoms, this is another embarrassing side effect of social anxiety disorder. The individual becomes intensely nervous and needs to race to the restroom, which in turn can cause the attention they desperately are seeking to avoid.

Increased Respiration Rates

Breathing rapid, shallow breaths can increase the anxiety that the individual is feeling, leading to a panic attack and shortness of breath. Deep breathing will alleviate the rapid breathing and stave off a panic attack usually.

Dizziness

Feeling faint, lightheaded, or dizzy can be caused by anxiety. Most likely, the combination of shallow breaths, racing heartbeat, and fear will cause this during an intense instance of anxiety.

Nausea Or Vomiting

Becoming nauseated is common with anxiety. Social anxiety sufferers can become physically ill over the anxiety they experience in social situations. The need to flee to the restroom draws unwanted attention to the individual, and the cycle continues with symptoms.

Intense Urge To Escape

In our lifetime, most of us will have the urge to escape a situation. With social anxiety, that urge is not something that dissipates over time. The individual who suffers from SAD has the urge that is so intense to escape that they begin to panic if they are not allowed to leave.

Negative Emotional Cycles That Cannot Be Controlled

The negative emotions a person experiences with social anxiety are overwhelming. These emotions cycle through the individual, leaving them feeling off-balanced and unsure of the next move. Due to these negative emotions, getting through the wall of anxiety is difficult, but not impossible.
The symptoms often debilitate an individual who suffers from a social anxiety disorder. Each aspect of their life is affected by the symptoms, and they find that life becomes too difficult to manage. This causes unfavorable reactions that alter the way an individual carries out daily tasks and encounters.
Such reactions include:

Complete Avoidance Of Social Interactions
Social interactions are healthy for us as humans. We need interaction with others to feel connected. Individuals who suffer from social anxiety disorder avoid these interactions at all costs. Usually, if an individual suffers a negative outcome, this will cause that individual to avoid the next one at all costs. Sometimes, imagining negative outcomes is enough to make an individual avoid socializing at all.

Avoiding Eating In Front Of Other People

Do others you know to avoid eating in front of people? Are you self-conscious about this activity? Since the threat of rejection, criticism, and judgment are constantly in the front of a social anxiety sufferer, this activity becomes too stressful to take part in because of the threat that they could be rejected.

Avoidance Of Public Speaking

When looking at the list of reactions and symptoms, it is understandable that individuals with SAD avoid public speaking. If they are required to speak in front of an audience, no matter the size, this can cause them to become physically ill. No one should doubt the realness of a social anxiety disorder; it is real, and people every day are affected by it.

Avoidance Of Writing In Public

Writing in public seems like a normal aspect of life, right? This activity can be crippling for social anxiety sufferers. Perhaps they had an instance as a child where they were teased for their writing habits, or they could have felt scrutiny from others for this activity. Whatever the reason, there is a reality for avoidance. Any undue and unwanted attention can cause these individuals to seek ways to avoid being critiqued as a means of survival.

Depression

Along with social avoidance, social anxiety disorder can lead to other mental health conditions. Depression can be caused by negative thought processes and the cycles of negative emotions. Once caught in this vicious cycle, it is hard for the individual to see a way of escape. Their life becomes determined by the choices and the coping mechanisms that they make or use each day.

Because of the increased likelihood of other mental health conditions, individuals who suffer from social anxiety should seek professional help along with self-help techniques. In combination, these two can work beautifully to restore a fully functional life for these individuals.

Prematurely Leaving An Event

An individual who enjoys certain activities will soon find that they will leave an even if they begin to feel uncomfortable. Using this coping strategy will keep them leaving an event earlier and earlier each time to avoid an uncomfortable outcome.

Focus Being Shifted Inward

SAD sufferers will begin focusing on themselves more and more, resulting in an attitude that could be deemed selfish. The reality is that these individuals may not be trying to be selfish, but they cannot help it. The attitude, to them, is self-preservation.

Constant Avoidance Of Attention

Along with the other types of avoidance, the avoidance of attention, in general, is a reality for SAD sufferers. Not wanting to draw attention to themselves keeps them from doing even the most basic things, including sneezing, blowing their nose, and using the restroom. If a gathering has happened begun, these individuals will avoid leaving, even if the situation becomes dire.

Keeping Silent

For some individuals, social anxiety keeps them from speaking at all. They have experienced intense anxiety over speaking, and this has resulted in them remaining silent. Even if it is not a constant behavior, such as selective mutism, social anxiety has caused enough of a change that these individuals do not feel comfortable speaking their opinion.

Not Looking At Other Individuals

Eye contact is uncomfortable sometimes, and for SAD sufferers, it can be debilitating. Direct eye contact can cause a person to feel scrutinized and uncomfortable. For social anxiety sufferers, this is unbearable and causes them to keep their eyes directed down or elsewhere. With the reactions to

social anxiety triggers, you can see that the coping mechanism put into place by these individuals can be more debilitating than helpful. It becomes difficult for these
individuals to become comfortable enough, or miserable enough, to seek out help. The cycle that they are caught in does not allow them to break free easily, so working towards a goal is going to be beneficial.
As you can see from the reaction list, an individual with social anxiety is going to suffer throughout their life without help. Essentially these individuals become isolated, which only continues a cycle of negative thought patterns.

Because of this, it's important that if you suffer from social anxiety, that you work with a mental health professional to overcome this disorder. Life is meant to be enjoyed and lived fully, and individuals who
Individuals suffering from social anxiety often have lost the hope that life can be more than what they are experiencing. They forget, or may not remember what life was like before they were disabled with doubt, fear, and more.

Triggers

When looking at the symptoms and the reactions to the symptoms, you might be thinking about what are the triggers that are common among individuals with social anxiety. This question is extremely important if you are suffering from social anxiety. You may feel alone in your battle, and you probably wonder if others are suffering as well. Remember to keep hope that by the end of this book, you will have a game plan to help you overcome the crippling effects of social anxiety disorder. While you may never be a social butterfly, you should be able to have the moral support that you need.

Triggers can include:

Being Introduced To New People

For individuals with social anxiety disorder, being introduced to new people can set off symptoms and reactions. The fear of rejection and judgment becomes intense and can affect other areas of their life. Understanding the difference between social anxiety and shyness in this instant is important. The shy person will lose any feelings of trepidation as they become more familiar with the new individual. An individual suffering from social anxiety will not become more comfortable in this new situation.

Being Criticized Or Teased

Criticism for social anxiety disorders can cause them to refrain from the activity that they have been criticized about because of the attachments been drawn to them. These individuals, even constructive feedback is a result of failure on their part.

Being The Center Of Attention

Being the center of attention is difficult for individuals with a social anxiety disorder. They do not do well with the direct focus and prefer to remain on the sidelines. Fear of attention was real and can cause anything from minor symptoms to a full-blown panic attack.

Being Watched Or Observed Performing A Task

This becomes difficult in the workplace where employees are regularly lodged for their performances. Social anxiety disorder sufferers feel that any mistake that they make is going to be detrimental to their outcome. While being watched, the individual will run negative scenarios and outcomes through their mind, often resulting in a self-fulfilling prophecy.

Having To Say Something In Public

Public speaking is not an activity for an individual with social anxiety. It's not a hopeless activity; however, with the proper tools and coping mechanisms, being a public speaker is entirely possible.

Most individuals do not enjoy this type of intense focus and would not choose this as a regular activity.

Meeting People With Authority

Authority figures cause anxiety and tension often even in individuals who do not have an anxiety disorder. When somebody has the power to control or correct your life and choices, this causes individuals to feel nervous in their presence. For social anxiety sufferers, meeting people with authority can cause them to go into full-blown panic mode.

Insecurity In Social Situations

Security is part of social anxiety. These individuals often feel out of place no matter where they are because, in their mind, they do not belong. Even if these individuals have been invited, they will still feel like they do not belong in social situations. They do not usually have the skills to be effective in a social situation, and therefore, they become nervous and easily panicked.

Becoming Embarrassed Easily

Any small attention, misstep, or mistake can cause embarrassment to social anxiety disorder. These individuals go their whole lives trying to avoid being noticed, and when they do make a mistake, their embarrassment is often more intense than it needs to be.

He's individuals will blush, tremble, and become short of breath as a response to being embarrassed.

Making Eye Contact

As mentioned before, my contact is difficult for most people, but for anxiety-ridden individuals, eye contact is nearly impossible. Forcing eye contact would cause negative reactions and often can cause full-blown panic attacks. The individual becomes worried about what the other person is looking at, thinking about, and judging them on which in return begins the cycle of negative emotions.

Swallowing, Writing, Talking Or Making Phone Calls In Public

For an individual who watches every step they make to avoid attention, confrontation, or judgment, the activities mentioned above are usually avoided and done only when they are alone. If that's impossible, these individuals will hide what they are doing from view. Consider being in a restaurant and needing to swallow, but not being able to do so for fear of criticism.

Dating

Dating can be stressful for many reasons. Usually, on that date, the two individuals will focus their attention on each other, often eat together, and required a verbal exchange. Consider doing this now with anxiety.

How does that change the scenario? Despite the desire for companionship and love, individuals who suffer from social anxiety find it difficult to enter the dating scene.
They would rather remain alone and go out on a limb.

Interactions With Extroverts

Extroverted individuals can cause severe anxiety and panic, and individuals with a social anxiety disorder. It becomes another situation to avoid. Extroverts tend to be chatty and overly avoiding, which in turn makes the introverted individual feel out of place.

Large Family Gatherings With Unfamiliar Individuals

Having anxiety around family members that you know is possible, but attending large gatherings with individuals that you are unfamiliar with is the norm. Being comfortable with an individual, however, does not mean that you will not experience the feelings of rejection and fear. It simply means that these feelings may become more intense and require leaving early.

Parties

Even if you are familiar with the host and the other attendees, parties can become difficult as well. The noise, large crowd, and excessive talking and cause you to feel self-conscious and out of place.

With social anxiety, these types of interactions are avoided, and if you would enjoy the event. With proper coping mechanisms, these types of events do not need to be avoided.

Unexpected Attempts To Initiate Conversation

Any attempt for a conversation that an individual with social anxiety was not anticipating can cause reactions. These conversations could be in line at the grocery store while waiting for a ride, or any other instances where an unknown individual will attempt to initiate conversation. These situations can be handled effectively and not cause stress.

Speaking On The Phone

For most individuals with social anxiety disorder, speaking on the phone causes negative reactions. This is especially true if the individual on the other end is unknown. Often speak on the phone can be too stressful so these individuals will opt for text messaging or emails.

Therapy And Group Sessions

The drawback to the avoidance tendency is that there be in group sessions often are avoided. This is unfortunate because individuals with social anxiety disorder need to be able to seek out help, and if being afraid to work with the therapist or the support group is impossible, the cycle will continue.

There are coping strategies for this as well, and we will get into it later in the book. Every individual with social anxiety makes life manageable at times for themselves.

These individuals learn coping mechanisms and other ways to make it through the day unnoticed. With the triggers mentioned above, these individuals often have at least 3-4 situations that will completely overwhelm them, making life more difficult for them.

This list can continue even past these triggers, but looking at the list, you can understand how an individual can become debilitated by social anxiety.

An individual who is suffering from social anxiety does not even realize that these are setting them off because the individual has more than likely not connected the dots. When you understand what triggers your responses, you are better equipped to handle them appropriately.

With that in mind, how can you learn to function with your social anxiety? What steps can you take, what treatment can you seek, and what alternatives are available to you? In the next chapters, we are going to delve in-depth with these topics.

Chapter 3: Distorted Thoughts: What Are They and How to Change Them

There are tendencies to succumb to distorted thoughts when you suffer from an anxiety disorder. How can you combat these thought processes, identify them, and relieve yourself from the false beliefs they allow?

Let's start by looking at the common distorted thoughts. From there, let's look at the ways to evaluate the beliefs that are entering into our heads.

Common Distorted Thoughts

A distorted thought is a way that our mind convinces us that something is untrue. This is usually revolving around the way we feel about ourselves in constant negative thought patterns.

These thoughts serve us by keeping us from seeing how inaccurate and irrational our thoughts are and is usually revolving around us feeling bad about ourselves. There are common cognitive distortions that happen in individuals.

We will explore these and their relation to social anxiety in this chapter. Hopefully, after seeing the 15 common thought distortions, you will be able to implement strategies to overcome them and identify them in your own life.

Mental Filtering

This is a thought distortion that filters out any of the positive parts of the situation. Usually, this is when an individual will focus on one negative aspect that keeps them stuck. Consider individuals suffering from social anxiety.

These individuals feel that everyone is judging them. In a social situation, consider individuals suffering from sad such as social anxiety disorder and someone says that the work this individual had done on a presentation was okay.

When using the fought distortion of mental filtering, this can be taken as criticism. Even if the statement of the presentation being okay were followed up with praise for other areas of the presentation, an individual who suffers from social anxiety disorder would morph that into dark tormenting thoughts.

Black-and-white Thinking

Black-and-white thinking does not allow for the nuances of real-life. It often services in individuals with a social anxiety disorder because they feel that everything needs to be completely perfect or else a failure. In this thinking, there is no gray area that does not reflect life appropriately.

When you have this type of polarized thinking, you are going to notice that everything is in either or. Either my presentation was fabulous, or it was an abysmal failure. An individual with social anxiety, this thought distortion can be crippling.

Constructive criticism is no longer working because they do not see the middle ground.

Overgeneralization

When looking at this type of thought distortion, this is when an individual is taking one single instance and making a conclusion based upon it.

A person who suffers from social anxiety is going to take each failed social attempt or one social attempt, and assume that each attempt at social interaction is going to be as disastrous as that one single moment in time.

They are going to continue to think that they are unable to interact with other individuals in this effectively causes these individuals to refrain from social interaction.

Jumping To Conclusions

Everyone is guilty of jumping to a conclusion at some point in their life. Jumping to conclusions means that you don't wait to see how a situation plays out, you use your judgment to determine beforehand how the situation resolves.

This can become problematic with an individual who suffers from social anxiety because their view of the world is already distorted. They take past experiences and conclude that that is how each experience is going to play out, causing undue stress.

When you anticipate how something is going to resolve, frequently you focus on the bad and refuse to look at the good.

Catastrophizing

When you expect the worst, and you believe that it any point in time the moment is going to end in disaster, this is what catastrophizing includes. For an individual with social anxiety, this is going to manifest when they look at a situation that makes them uncomfortable.

As you know, this could be any situation that involves socializing. Expecting the worst for interactions is common with individuals suffering from social anxiety. There are strategies that we will discuss later to help curb this type of thinking.

Personalization

Believing that everything that somebody else does or says is some personal reaction to yourself is personalization. This can be something as simple as seeing a post on social media and assuming that the individual is targeting you. For individuals with social anxiety, personalization can cripple you.

When you believe that everything is directed personally at you, it becomes impossible to separate truth and reality from the distortion that you are experiencing. Because of the high-level of personalization, social anxiety sufferers confine that interacting at all is not an option, and they choose to refrain from interactions completely.

Control Fallacies

This belief involves external control and internal control. External control is where we feel that we are just a victim of destiny. For a person with social anxiety, this is going to manifest by feeling that everybody is out to get you and that nothing that you do is going to be good enough as a result. Internal control is where you take on responsibility for everyone and everything around you.

A good example is when your spouse is unhappy, and you assume that it's because of something that you've done. These two different types of something that you've done. These two different types of control that will cripple an individual with social anxiety because these individuals already believe that nothing, they do is good enough and that they are responsible for the unhappiness around them.

Fallacy Of Fairness

Individuals who suffer from this thought distortion believe that everything is unfair because it does not agree with their idea of fairness. I'm sure you've heard that life is not always fair while you were growing up, the people who believe this thought distortion throughout their life are resentful, angry, and hopeless because they need life to be fair and things to work out in their favor.

Not understanding that life does not work out this way is difficult for individuals to accept and understand.

Blaming

Blaming can take on two forms: blaming everyone else or blaming only themselves. An individual with a social anxiety disorder will often blame themselves instead of others. Consider an individual who feels responsible for every bad thing that has happened to another person, such as a spouse or partner. Internalizing the blame will cause the cycle of negative thinking to circulate, never letting the sufferer off.

Should've

Saying things like: I should have done this better, "I should do that more" can make you down on yourself and cause thought distortions. An individual with SAD feels this on a personal level. They take everything that they feel they have done wrong and internalize this, which causes the individual to repeat a negative thought cycle.

Emotional Reasoning

Believing that the emotions are true and accurate describes emotional reasoning. "If I feel like a failure, it must be true." This distorted thinking is problematic because when you follow your emotions, you will find that it is harder to follow what you believe.

The emotional reasoning distortion causes problems with social anxiety sufferers because often these individuals rely too much on their emotions and connect them to their perceived failure.

Fallacy of Change

If someone is pressured to change enough, they will change: that is the idea behind this distorted thought. This type of distorted thought process will usually revolve around relationships. Usually, one or the other partner thinks they need to change the other one for one reason or another.

Global Labeling

Taking one or two negative qualities and making general assumptions about another person or themselves. This is another common thought distortion for social anxiety. With global labeling, an individual will judge an entire situation based on the task that they have failed. Consider the example of attaching a negative connotation to someone based on their behavior.

The term jerk is usually attached to an individual who annoys someone once, and as a result, this person is labeled this from then on. The language used in global labeling is highly skewed and emotionally charged, which causes an individual to attach in unhealthy labels to any of themselves or someone else.

Always Being Right

This distorted thinking is when an individual has to show that they are right all the time and that their actions are the only correct ones. They do not understand that there is distortion in their thinking and they go to great lengths to demonstrate that they are correct and everybody else is wrong. Often they will not even consider the feelings of another individual.

The Heaven's Reward Fallacy

This thought distortion is where you believe that the sacrifice that you have made is going to pay off in the end. You believe that there's some scorekeeper somewhere that is going to make sure that in the end everything is fair and you are the one along with others who work hard are going to receive the largest reward. If you suffer from this, you're going to feel ripped off if you do not receive the reward that you expect in the end.

Combat Thought Distortions

These common thought distortions make it difficult if you are suffering from social anxiety. Social anxiety itself causes your thought processes to be skewed and difficult to handle. How do you fix these distorted thoughts? In the next part of this chapter, we are going to look at the ten different ways that you can begin to fix the thought distortions.

Begin By Identifying The Thought Distortion

This is the most important step in the process of fixing thought distortions. You need to identify which distortion you are experiencing and how it is affecting your life. You need to identify and track these distortions in your daily thinking before you can begin to start working to change the distortions.

Creating a list of any troublesome thoughts that you could be having throughout the day is a great place to begin because as you write them down on a regular basis, you're going to see any patterns. When you look at the thought distortions, and you see a pattern, you are going to see which type of distortion you are predisposed to. This type of activity requires that you keep a journal with you at all times, and you write down the thoughts as they come.

Remove Yourself Emotionally

It's important to look at the evidence in front of you so that you can examine it and find out which part is rational. When you look at the situation objectively, you are going to allow yourself to identify the basis of your thought processes. If you continue to look at the situation where you with an overly critical eye, keep a list of past ways that you've been successful. By looking at this list, you are going to realize that you're not doomed to failure and that this itself is the rational thought process.

Keep in mind to use facts rather than opinions. A fact is something that you can solidly identify, such as I forgot to wash the dishes. A statement such as I am an idiot is an opinion. Understanding the difference between the two is going to give you success in determining your thought processes.

Look For Double Standards

We oftentimes hold ourselves to a higher standard than we would someone else. The way that we talk to herself and the inner chatter that we experience is often more harsh and severe than we would talk to someone else. When you talk to someone that you love, you would never consider speaking to them the way you allow yourself to speak inside your own mind.

You need to learn to treat yourself with the standard that you would treat everyone else. When you refuse to do this, you are engaging in a double standard that does not give you the same encouragement that you deserve. Imagine if you were working on a project with a friend, and told them that they were going to screw that project up just like they scrub everything else that they work on.

That's something that you would save yourself, but would you ever consider saying that to a friend? The chances are that you wouldn't, which proves the point that you need to break those negative thought processes and offer a more rational response to self.

Allow Gray Areas

When you do not have areas of gray in your thinking, you are doing yourself a disservice. Remember, this is called polarized thinking and can result in overly critical self-talk. There is no either-or situation, but rather there is a scale that varies. When you evaluate things, it's important to scale it on a 1 to 10 basis and to look at the experience in the process rather than just the outcome. Do not allow yourself to fall victim to the polarized thought distortion.

Try an Experiment

With this method, you need to take the irrational thoughts that you are having and balance them against facts. For a person with social anxiety, this could be difficult.

Take A Survey

This method requires that the individual who is having distorted thought processes asked others who have been in a similar situation. When you talk to trusted sources about their experiences, it's gonna make it easier for you to understand the irrational path that your thoughts are taking because when you seek the opinions of others you're gonna have something to gauge what is realistic and what is not realistic.

Keep in mind to ask people who are not suffering from an anxiety disorder something such as that because you want to be able to look at something that's not a distorted thought.

It's also important to ask several different people so that you can gauge what reality is and what is simply distortion. The opinions and experiences of others that you trust can be helpful, especially to somebody with a social anxiety disorder.

The Semantic Method

When looking at this method, it's looking at how people have unwritten rules about their behavior, but these rules do not make sense. Other people find these rules to be stifling and judgmental in regards to an individual, and most people can see these sets of rules as being unhelpful or hurtful.

When you use I should statement, it makes it difficult not to follow that distorted thought process. If you're saying I should not make people angry, then you are giving yourself an unrealistic expectation to live up to in which causes a vicious cycle of
distorted thoughts.

When looking at the statements, reframing the statements to wouldn't it be nice if is a lot better than saying I should. Consider wouldn't it be nice if I didn't jump to conclusions about others? Or wouldn't it be nice if I exercise more? The statements are empowering rather than hurtful, which is going to help a person who is recognizing their distorted thoughts.

An individual suffering from social anxiety is gonna find this particularly helpful because it allows you or the individual to step back and realize that life is not definite, and the way that we talk to ourselves and the expectations that we hold ourselves up to are not definite as well.

Definitions

Do you like to argue? This method is perfect for individuals who like to argue and want to argue about even the smallest details. This requires that you attach a definition to the labels that you are putting upon yourself.

When you look at the meaning of the labels, the specific behaviors that are attached to them, or even the behavioral patterns, then you are going to notice that that's not you. This is particularly helpful in combating the global labeling thought distortion process.

Giving a broad label based on just a few characteristics causes the thought processes to become distorted and unrealistic. Consider using this method to define instead what these labels actually mean. For an individual with social anxiety, this is gonna be helpful because when you have a solid definition in front of you, it's easier to not contribute those characteristics to yourself.

Be careful that while doing this, you are not actually looking at the list and checking off all of those characteristics as to who you are because that's being counterproductive.

Re-attribution

The re-attribution method is particularly helpful for combating the personalization and blaming distortions. Remember with these two particular thought distortions an individual blames themselves for any of the negativity experienced in life, despite where the negativity was actually caused.

You're taking the external factors, and the actions or contributions from other individuals and you are placing that on yourself. When you reorganize the responsibility and place that responsibility where it belongs, you are going to experience freedom from these thought processes.

It's easy to get caught up in a cycle of blaming yourself but doing so causes detrimental effects on our self-esteem. The reality is that most people are not to blame for everything that happens, but as humans, we tend to take on the responsibility.

This is particularly seen in individuals with a social anxiety disorder. Your mind is telling you that everything is your fault, even when in reality this is not true.

Cost To Benefit Analysis

With this method, you are going to tackle your irrational beliefs and counteract them with facts. Make a list of advantages and disadvantages for any of the feelings, behavior, or thoughts that you are experiencing.

This is a good way to look in black-and-white to what this distorted thinking is costing you. When you look at the pros and cons of these thoughts, you're going to have the chance to fight back and reclaim your life.

The distorted thoughts that an individual feels can take a toll on their life. In particular, individuals with social anxiety disorder find it difficult to function because their thoughts are telling them that people are judging them; they are not good enough, and that they are failures.

The intense feelings of individuals with social anxiety disorder find it difficult to function in life because of these distorted thoughts. It may take time to work through all of the distorted thought processes that you are experiencing, but the time and effort are going to make your life that much better.

Consider to take the time, journal, and discuss with people that you trust how you are combating these thoughts, and how they are going to benefit you in the long run.

No one wants to be miserable, and the first step to combating irrational thoughts is to begin. Take action and reclaim your life.

Chapter 4: Positive Thinking

Positive thinking is a mental attitude that you adopt where do you expect nothing but good and favorable of comes in life. This requires that you change your thought processes to create thoughts that are going to transform your energy into your reality. By adopting a positive mindset, you are waiting and expecting happiness, health, and a happy ending.

This is not a rosy outlook on life necessarily, but rather it is you using your mental energy to attract and manifest happiness and success.

You are going to see the benefits of positive thinking. Mental health is going to improve because positive thinking is going to eliminate paranoia, negative thought processes, or distorted thoughts. These can lead to stress and depression, which is going to hurt your health should you continue living with these two. Individuals who adopt positive mindsets are mentally and physically healthier, happier, and experience more content in their life.

Perhaps you're asking yourself if positive thinking actually works, and it does.

Research has been done on the benefits of positive thinking, and they are revealing that it's more than just being happy or upbeat.
These positive thoughts that you adopt are going to benefit your life by creating value. You are going to smile more, but that's not the main benefit.
When you adopt a positive mindset, you are empowering yourself to add value to your life by building different skills that are going to benefit you in the long run.

Remember that this is a remarkable strategy to put into effect in your life. If you happen to know somebody who is continually positive, you were going to notice that they seem to have better outcomes in their life. Remember that your mind is powerful and that you can change what your world looks like with your mind.

When individuals are healthier and happier, they are able to get what they want out of life, and they are able to know how to go about it. Remember that adopting a positive attitude can change your life dramatically. Are you ready for that change? If you are, let's get started.

Positive thinking can help you overcome anxiety, worry, and depression. By simply changing your mindset, you can shape your world and your future into something better than you can imagine. Suffering from social anxiety can cause you to think that your world is your reality, and there's nothing you can do about it.

This is a common distorted thought and as such, is entirely inaccurate. You are in control of how your world is and what your future is as well. Positive thinking and adopting a positive mindset are going to help you change and shape your destiny.

Alleviating Worry

Individuals, as humans, are going to worry about certain situations. Worry is a part of life, but when you worry about everything and anything, this then becomes a problem. Focusing only on the negative is going to harm you mentally and physically. Individuals who suffer from chronic worrying and anxiety find that they can only focus on the negative, and seeing the positive in a situation is impossible. You become fixated on these negative scenarios and can end up self-sabotaging.

Allowing yourself the time and energy to make an effort to refocus your thought processes is possible. When you fixate on positive outcomes rather than the negative counterparts, you are going to find that the positive outcomes frequently happen in the negative outcomes are less frequent. This is manifesting your destiny, and this is possible by simply reshaping your thoughts.

Keep in mind how important your thoughts are.
You can remove yourself from the negativity of others throughout the day, but when your thoughts are consistently negative, there's no way to escape that.

This becomes particularly troublesome for individuals with a social anxiety disorder. The constant worry that these individuals experience causes them to focus solely on the bad, and they are unable to see the good.

With that in mind, how can you use positive thinking to overcome worry? You need first to find something that is going to help you be consistent. This requires quite a bit of focus, but in the end, it will be worth it. Earlier in this book, we talked about recording your irrational thoughts and combating them by reshaping those thoughts.

Changing your worrying mindset to a positive mindset is as simple as recording your negative thoughts and underneath those negative thoughts, rewriting them to focus on the positive. The more you do this, the easier it becomes and the less negativity you will experience in your thought processes.

Consider the thought:

"If I walk out the door today, I'm going to be struck by a bus."

"When I walk out the door today, I am going to experience happiness and laughter."

By looking at the first example, you can see the worry and negativity in that simple statement. We shape that first statement with a positive aspect and thus take one more step to change a negative into a positive.

Alleviating Anxiety

Anxiety can cripple an individual. Positive thinking is going to help you overcome your anxiety because anxiety at its core is negative thinking. Similar to worry, which we discussed above, overcoming anxiety is as simple as changing your thought processes. To overcome anxious thinking, you need to stop your negative thoughts when they happen. Each time you have a negative thought, it's important to reshape it while looking for the positive.

As with the technique for worrying, a thought diary is a good idea. When you record your thoughts, and then you record them in a positive light, you're going to experience relief. The relief comes from releasing the negative thinking and embracing the positive. This should also be done with any irrational thought processes or distorted thoughts that you are having because the key is to identify the thought as it happens. When you finally are able to identify these thoughts at the moment, you're going to see that they come less frequently and that you are more positive in general.

Remember that positive outcomes are possible and that expecting only the worst is simply worry anxiety, and distorted thinking. Do not allow yourself to dwell on these areas because your physical and mental health is at stake.

Alleviating Depression

The key to overcoming depression lies with understanding first how depression begins. Depression can be a chemical imbalance in the brain, or it can be situational. Depression usually begins with a series of thoughts that focus on the negative and become a habit or a different way of thinking. At this point, depression has taken over, and you can see the effects rapidly.

Along with the mental health side effects, the physical side effects of depression are just as alarming. Weight gain or weight loss, fatigue, muscle pain, joint pain, and more are all common physical side effects of depression. When these side effects happen, often they continue to leave the individual on a vicious physical cycle that's hard to break. Physically the individuals stop having the will to do what they love, which in turn leads to physical complications as well as the mental health complications.

You may not even realize that your thought processes had changed from positive to negative. Even if you happened to be in an upbeat and positive person in the past, the quick replacement of negative thought patterns happens rapidly. You need to replace those thought patterns and change them as quickly as possible to overcome depression.

The trouble lies in the cycle that a depressed person is caught in because once you feel sad and depressed, it's difficult even to want to change the negative thoughts to positive ones. Your mind becomes comfortable with these thought patterns in your mind wants to hang onto them. The importance of ridding your mind of these negative thoughts cannot be overstated. If you continue with these negative streams of thought, they are going to become your police system which is going to alter not only how you feel the world but also your personality.

Once you allow these thought patterns to become part of your belief system, your subconscious will start to create instances that mirror your negative thought patterns. This is a form of self-sabotage which is common with depressed thinking. The further and longer you hang onto these depressing thoughts, the more difficult it's going to be to overcome.

The first step is to change your thoughts. You need to make up your mind and push your mind to come up with positive thoughts to counteract the -ones. This is like we discussed before; you need to counteract the negative thoughts with a positive thought. As you do this step-by-step, you are slowly going to find that it's easier to be positive rather than negative. As a result, you are slowly going to make your way out of the depressive state.

If you feel that you are not able to push yourself and reframe your thoughts, this is simply because your energy is low due to the depression. The important idea to remember is to start with one thought and continue from there. During the process, make sure that you are encouraging yourself to overcome this negative thinking.

On her good moments, post encouraging notes around the house, at the office, and even in your vehicle so that you see them even when you feel hopeless. Make sure these notes are encouraging and positive.

Remember that overcoming anxiety, worry, and depression is a long process. The process is by no means a simple fix, but that just makes the rewards much sweeter.

Steps To Adopting A Positive Mindset

Now that we've discussed how a positive mindset can benefit you let's look at some practical steps that you couldn't enact in your life to make this a possibility in reality.

Remember that a positive mindset is going to set you up for success in overcoming worry, anxiety, depression, and social anxiety.

If you suffer from any of these, you understand the importance of finding a solution. Below you will find ways that are going to help you begin reshaping your thought processes.

Positive Affirmations

Do you know what positive affirmations are? Positive affirmations or statements that leave you feeling empowered and ready to face the day. You should repeat affirmations in the morning before you begin your day and also in the evening before you retire for the night.

Affirmations focus on the good and do not allow you to focus on the negativity of life. By using positive affirmations, you are empowering yourself to change your thought processes.

Here is a list of positive affirmations that can help you change your thinking and improve your outlook on life.

- I am in charge of how I feel today.
- Today I choose happiness because I can.
- I can overcome obstacles today.
- I am more than my past, and my future is bright.
- I am brave.
- Today I will be successful.
- I am peaceful, and I choose peace.

These are just some examples of positive affirmations that you can repeat to yourself in the morning and in the evening. By focusing on what you are that is positive, you are allowing yourself to reframe and eat your rational thoughts that you may be having because you are counteracting those thoughts. The power of positive affirmations is not to be scoffed at because your thoughts are powerful enough to change who you are and where you are going. How amazing is that? You alone can change your outcomes by changing your thoughts.

Focus On The Good

There is no such thing as too small of a success. Each when you experience in life is important. If you are working to overcome anxiety, worry, and depression each day that you experience positivity is going to be monumental.

Learn to embrace these moments and celebrate them because they are important. Remember that there's nothing that you do that is too small to show progress.

Progress comes from the small steps, not the gigantic leaps because each day as you take smaller steps towards a happier future, you are going to see
progress. Remember that you will loan control how you view life and as a result, you can celebrate.

Find Humor In Every Situation

Laughter is powerful. Simply laughing in a bad situation is going to alleviate your stress, anxiety, and fear. The power of humor and laughter will surprise you. When you're feeling anxious during that situation, find something that will make you laugh.Even if you look like you've lost your mind, laughing is going to release the stress that you are feeling.

Try to focus on finding humor or laughter as well as positive in each situation and see how your world begins to change for the better. Try to focus on finding humor or laughter as well as positive in each situation and see how your world begins to change for the better. You'll find that it is easier to tolerate the bad when you are also making a point to look for the good. By finding a positive in each situation, you are reassuring yourself that it is not as bad as you may be thinking.

This takes a conscious effort, but in the end, it is worth it because you are able to learn how to defuse a potentially disastrous situation. By disastrous, I mean that you are defusing the situation that could lead you on the spiral down the path of negativity and install any positive progress you have been making. Choose positivity, humor, and laughter and embrace these to relieve stress, anxiety, and worry.

Failures Are Lessons

The simple act of reframing how you see obstacles or failures will also reshape your thought processes. When you make the decision to find the lesson in failure or obstacle, you are really making the decision to learn and move forward on your journey. Obstacles are nothing more than a learning opportunity, and each failure that you experience is a way to learn and grow.

By doing this, you are going to see that progress will be made and that having a misstep is not the end of the world. Remember that it obstacle or failure is not a stop sign, but are rather yield signs in life. Take a moment to look around when you come across these yield signs and learn.

Quiet Negative Inner Chatter

When you have negative self-talk, you may have noticed that your outlook on life is negative as well. If you haven't noticed this, it could be because you are so entrenched in the negative thoughts that you are unable to see this pattern. When you learn to quiet the negative inner chatter, you will see relief. Adopting a positive inner chatter is an important step towards changing your reality.

Practice Mindfulness

By learning to remain in the present, you are refusing to rehash the past or worrying about the future. When you worry about the past, it becomes difficult to maintain positive thoughts.

The same is true when you can't stop worrying about the future. Your thoughts become caught up in memories and worries that you cannot control at the moment, and this leads to a vicious cycle of worry, anxiety, and depression.

Mindfulness means that you are going to remain in the present. Practicing mindfulness is a simple exercise in reshaping your thoughts and focusing your mind. This is a conscious effort that you need to make on a continual basis, but in the end, the rewards are great.

Once you stop worrying about the future and thinking about the past, you are free to begin your journey in the present. This is one of the biggest steps that you can take towards positive thinking.

Surround Yourself With Positivity

Who you surround yourself with and what you surround yourself with matters. Do not negate the importance of positive people. If your friends are negative people, this is going to bring you down.

Surrounding yourself with people that are upbeat and look on the bright side is also going to reshape your thinking because it's hard to be negative when positive thoughts are basically slapping you in the face.

Also consider if you are the individual in the group that is negative, you could be the reason somebody else is being brought down.

This is nothing to feel bad about because at some point in our lives we all are that person, but trying to change your own attitude is going to also affect the others around you.

Make sure that you are also listening to positive materials in music, reading positive items, and absorbing information that is positive. You would be surprised at what happens to your mood and your mindset when you take steps to eliminate negativity that surrounding you. Perhaps you are not even aware that you are absorbing this negativity.

Show Gratitude

How often do you practice gratitude? If you have lost track of expressing what you are grateful for, now is a good time to begin again. The effects of gratitude on your mindset are also astronomical. Focusing on the areas in life that you are grateful for allows you no time to focus on the areas that are causing you concern and worry.

Consider starting a gratitude list or gratitude journal. On a daily basis, right on five things that you are grateful for because when you do this, you are reminding yourself that good, positive things did happen throughout the day. When you feel especially dejected and despondent, or worry and anxiety are taking over, take out your gratitude journal and focus on the good areas that you may have forgotten.

Smile

When you smile, your mood automatically lifts. Smiling leaves you no time to focus on negativity because just that simple act is going to brighten your day.

Even if you do not feel like smiling, try anyhow. When you force yourself to smile, try finding a positive idea or event to smile about because then you can transform a fake smile into a genuine one. Doing this will also help alleviate worry and anxiety.

Never underestimate the power of a smile because not only does it brighten your day but it could also brighten somebody else's day as well.

Know Your Mission

What do you feel your mission is in life? Have you considered this at all? If you have not considered what your purpose is in life, take some time and look into it. Try understanding why you are here and what you need to be doing because giving yourself a purpose is going to change your mindset as well.

People who suffer from anxiety, worry, depression have a hard time looking at what they need to accomplish because of the cycle of negativity spiraling around in their minds.

If you are journaling, dedicate a section to figuring out what you like to do, what you are good at, and how these can converge into one purpose.

If you are journaling, dedicate a section to figuring out what you like to do, what you are good at, and how these can converge into one purpose. Giving yourself a mission in life is going to make a difference. A mission loses positivity, and from there, you can continue the process of adopting a positive mindset.

Use Positive Words

When you describe your life or your situation, using positive messaging is going to make a difference. Instead of saying,

“I'm terrible at,” consider saying, “I may not be good at this, but I am good at that.” Another way to do this would say, “I have a good friend. I have what I need.” Focus on what you have and not what you don't have, and use positive words to describe your situation. Even a simple change of the wording that we use is going to make the difference.

When you practice your affirmations in the morning, and the evening, this is a good way to work into the habit of using positive messaging.

Positive messaging is going to take you from feeling worried, anxious, and depressed and help you develop that positive mindset that you are striving for because to attract positivity, and you need to give off positivity. You do this by changing your word choice.

A positive mindset is going to make all the difference in the world when you are looking at changing the way you deal with anxiety, stress, worry, depression. Positivity is vital to overcoming negative thought processes.

Try the steps above, and see how they fit into your life. Remember that a positive mindset does not come in a day, but it is something that you need to work on every day.

It's not automatic, but rather it is a gradual process that you are going to see results with because when you let go of the negativity, you are going to find that you may have even forgotten that it left.

Conclusion

Thank you for making it through to the end of *Social Anxiety and Shyness*, let's hope it was informative and able to provide you with all of the tools you need to achieve your goals whatever they may be.

The next step is to:

- Learn the causes of social anxiety disorder
- Begin a journal to track your negative thought patterns
- Look into finding a mental health professional, if you have not already
- Identify the steps you can take to practice mindfulness
- Look at the steps involved in adopting a positive attitude
- Identify your triggers and reactions to social situations
- Take time to identify distorted thought processes
- Look at the situations that you avoid because of these distortions
- Look closely at what you fear
- Practice relaxation techniques and exercises
- Adapt these steps to fit your personal social anxiety
- Remember that not everyone has the same symptoms, reactions, or causes, and that is fine!

Social anxiety is a curable mental health disorder that most people do not seek help for but should. Congratulations on becoming a step closer to living and thriving in the life that you were meant to live! Good luck on your journey, and may you find much success!